Catiele Bruna Dell'Osbel

Alternative periodic table using recycled paper

Catiele Bruna Dell'Osbel

Alternative periodic table using recycled paper

A different alternative for learning

ScienciaScripts

Imprint

Cover image: www.ingimage.com

This book is a translation from the original published under ISBN 978-3-8417-0784-0.

Publisher:
Sciencia Scripts
is a trademark of
Dodo Books Indian Ocean Ltd. and OmniScriptum S.R.L publishing group

120 High Road, East Finchley, London, N2 9ED, United Kingdom
Str. Armeneasca 28/1, office 1, Chisinau MD-2012, Republic of Moldova, Europe
Managing Directors: Ieva Konstantinova, Victoria Ursu
info@omniscriptum.com

Printed at: see last page
ISBN: 978-620-8-60673-2

To my parents Aureo and Claci, my brother João Vitor and my love Vanderlei, who have encouraged me throughout this journey. With their love, affection and attention, they have never let me down, being the essential support for this dream to come true, for the simple fact that they believe in me!

To my loves...

DEDICATE

ACKNOWLEDGEMENTS

Firstly, I thank GOD, to whom I owe everything I am.

To my supervisor, Professor Sandra Inês Adams Angnes, for her patience, her suggestions, for believing in this work and trusting in my ideals.

To the professors, especially my co-supervisor Professor Edneia Durli, colleagues and all the members of the Chemistry Degree Course, who directly or indirectly contributed to the completion of this work.

To my parents, who have interceded with GOD for my success and happiness at every difficult moment in my life.

To the most special person in the world, who has patiently accompanied me along this path, for all the love, affection, understanding and encouragement, for the moments of anguish and worry caused by me, for my absences during the realisation of this work, I dedicate this achievement to you with gratitude and love.

To my brother João Vitor, who often missed me because he wasn't in the family, because of the fights and arguments.

To my eternal companions who spent four years together, for the friendships we lost along the way and the ones we managed to end together.

"Each of us makes up our own story, each being carries **the gift of being able to be happy..."**

(Almir Sater)

SUMMARY

The aim of this work was to develop a different form of teaching practice for the study of the Periodic Table with students from a special Correction of Flow series at the Professor Anacleto Damiani Basic Education School in the town of Abelardo Luz, Santa Catarina. The Alternative Table was handmade by the teacher with recycled paper, produced by the Flow Correction students, by cutting, assembling and writing. A total of 118 cubes representing the elements of the Periodic Table were made, with information on each side, such as the symbol and name of the chemical element, atomic number, electronic distribution, mass number and applications. The Periodic Table was used in Science and Chemistry lessons, during which topics contextualised to the environment and students' daily lives were covered, such as preservation, rubbish, composition, recycling and reuse, relating some elements and chemical substances to the elements of the Periodic Table. The specific study of the Periodic Table covered the organisation of chemical elements into groups and periods based on their electronic distribution, exploring periodic properties and the study of covalent and ionic chemical bonds. After the lecture and solving exercises, a questionnaire was administered on the content worked on during the lessons, dividing the class into two groups. Group A solved the proposed activities using the Traditional Periodic Table and Group B used the Alternative Periodic Table. It was observed that the Flow Correction students who used the Alternative Table were more motivated and interested in the object of study, as well as solving some exercises with less difficulty, when compared to the results obtained when solving the activities proposed by the students who also used the conventional Table, probably due to the fact that the Alternative Table can be manipulated during the exercises, as it comes with buttons, sticks, arrows, cubes and the electronic distribution, which made it easier to solve the proposed work, since in a representative way the student can visualise and propose the compounds suggested in the exercises.

Key words: Learning. Chemistry teaching. Periodic table. Correction of Flow.

SUMMARY

CHAPTER 1

INTRODUCTION

> The students in the Correção de Fluxo school programme have age and grade/year gaps in primary and secondary education, which is one of the major challenges to be overcome by the Department of Education. In this sense, LDB 9.394/96 in its Article 24, item V, paragraph b, supports the adoption by education systems of actions that make it possible to accelerate studies for students who are behind in school (BRASILIA, 2011, p. 4).

For students to be eligible for the Flow Correction programme, they must be very old and in 5th, 6th or 7th grade classes, with no prospect of finishing primary school.

The Professor Anacleto Damiani Basic Education School has a Flow Correction class and one of the aims is for these students who are in an age-grade gap to have the opportunity to complete primary and secondary school, as well as regularising the age-grade gap of students enrolled in primary and secondary school in pedagogical activities. These students have a high degree of difficulty and for this reason this work aims to demonstrate a teaching methodology in the subject of chemical sciences that involves the student in the learning process in a participatory, creative and stimulating way, making the student seek knowledge of their own volition and above all build their own intellectual structures, based on group work, contextualised with the use of easy-to-use teaching material.

Concerned about learning and also in an attempt to motivate the participation of the students at the Professor Anacleto Damiani School of Basic Education, we used a contextual approach related to the environment through selective collection and recycling of paper, taking advantage of this to make an Alternative Periodic Table, proposed by the teacher and used by the students of Flow Correction, to study the properties of chemical elements and chemical bonds, thus seeking to make the students active subjects in the learning process.

According to Campbell (1989), the use of an Alternative Table is an important resource for developing thinking skills such as: imagination, interpretation, decision making, creativity, hypothesising new situations that arise, thus stimulating the skills and construction of the

knowledge. That's because, normally, in the classroom, the student is faced with a board: The Periodic Table of the elements. What do they see? Perhaps they see several rectangular boxes arranged in columns, where each box contains alphabetical symbols and some numbers without much meaning.

> It is therefore believed that in primary school, students should have a general understanding of certain properties, such as the classification of elements into metals, semi-metals and ametals. They are also expected to have a grasp of the classification of families, essentially the representative groups and the periods of the periodic classification. This basic reading of the Periodic Table, as well as the interpretation of chemical symbols, the organisation in ascending order of atomic number and the symbolic writing of elements and some compounds is fundamental preparation for secondary school (TASCA, 2006).

Based on these observations, the aim of this work is to motivate the participation of the students of the Professor Anacleto Damiani Basic Education School in chemistry lessons by studying the elements of the Periodic Table using an Alternative Table made by the teacher from recycled paper produced by the students of this school during science lessons.

1.1 Problem

Recycling paper using a contextual approach related to the environment with 8th grade students at the Professor Anacleto Damiani Basic Education School, using this paper to make an alternative Periodic Table for use in Science - Chemistry classes, will make students more interested in the object of study? Participate more in lessons? Favour the appropriation of knowledge?

Will using the Alternative Table bring better results than using the Traditional Table?

1.2 Hypothesis

Contextualising chemistry teaching with the student's everyday life, addressing relevant issues such as environmental preservation, the production of handmade paper and using this paper to make an Alternative Periodic Table for use in science and chemistry classes, gets the student involved and more interested in the teaching process, taking ownership of knowledge.

It is believed that the Alternative Periodic Table will be a component to help in the classroom, enabling the teacher to get to know their students and the reality of the group better.

1.3 OBJECTIVES

1.3.1 General objectives

To develop a differentiated form of teaching practice for the study of the Periodic Table with students from a special Correction of Flow series at the Professor Anacleto Damiani Basic Education School in the western region of Santa Catarina.

1.3.2 Specific objectives

C Selectively collect paper;

R Recycle coloured paper by hand through group work;

C Make an alternative Periodic Table to study the chemical elements, physical properties and chemical properties;

A To assess the students' interest in studying the Alternative Periodic Table and compare it with their knowledge of the Traditional Periodic Table.

CHAPTER 2

DEVELOPMENT

2.1 MATERIALS AND METHODS

2.1.1 Materials

The materials used to make the recycled paper were: cloths, a blender, large basins and buckets (various), artificial dyes, water, a ruler, pencils, coloured pens, used paper, sieves, scissors, contact paper, a 30cmx60cm wooden box.

2.1.2 Methods

To develop the work and collect data on the subject, we used experimental practice and group work to produce recycled paper with the students and the construction of a representational teaching model of the periodic table by the teacher using recycled paper. Bibliographical research and field research were used to support the study, using a questionnaire as a data collection tool to explore interest in and understanding of chemistry and specific chemistry content, with the aim of students understanding periodic and aperiodic properties, including how the elements were arranged in the table and how these properties relate to the formation of substances based on the study of covalent and ionic bonds.

2.1.3 Recycling paper crafts

The methodology available at (http://www.fazendapassaredo.com/como-reciclar-papel.html) was used to recycle paper by hand. The students went round the classrooms following the instructions given by the science teacher, talking about the importance of separating rubbish and recycling, dealing with environmental issues such as the life expectancy of plastic, glass and rubber, as well as the consequences caused to the environment when these materials are disposed of indiscriminately.

Bins for sorting rubbish and collecting waste paper were placed in approximately seven classrooms. The collection bins were left at the beginning of the morning (in the first class) and removed at the end of the fifth class.

Flowchart 1 shows the production stages of the recycled paper produced by the teacher and students during science lessons.

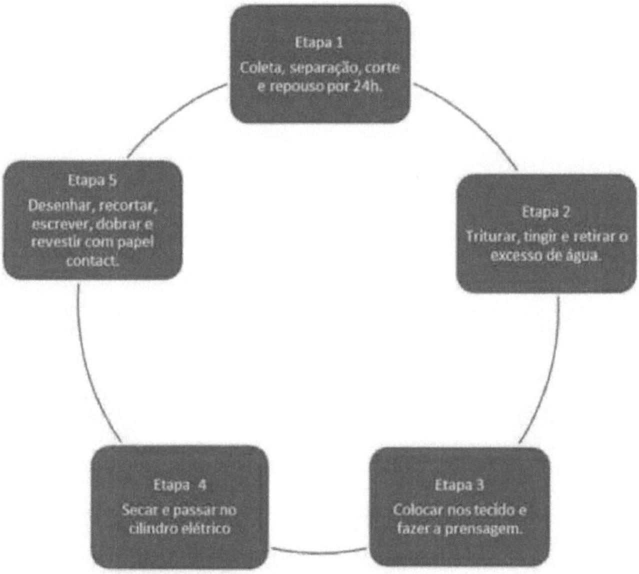

Flowchart 1: Recycled paper production stage.

According to the sequence shown in the flowchart above, the paper recycling stages consist of:

Cut the paper into small pieces, place in a container, cover completely with water and add colouring. The higher the water temperature, the faster the paper softens. The paper fibres can be coloured with dyes, which can be found in markets or dye shops;

Once the paper has been homogenised, excess water must be removed using sieves;

To make the paper thinner, it is first compressed with the hands and then covered with fabric. The paper is pressed using boards and weights (bricks, for example).After 24 hours, the paper is slowly removed from the press and the fabric covers are removed.

drying

It is first done in the shade and then dried in the sun;

As the sheet of paper dries, it will gradually move, so be very careful when removing the fabric to avoid breaking the paper. After the paper has dried completely, it is passed through the electric roller to make it thinner and more uniform, correcting imperfections;

P Finally, paper can be used for cut-outs, drawings, folding, etc...

2.1.4 Making an Alternative Periodic Table

To make the Alternative Periodic Table, handmade paper was used, as shown in the steps below:

1ª Stage

Using the methodology proposed by Júnior Bairro (2012), with the help of a cube model, the strokes needed to form the cube are drawn with graphite on recycled paper, measuring 2.8 cm X 2.8 cm for each face, as shown in figure 1.

Figura 1- Folding model used to make the cubes to represent the elements of the periodic table.

Source: Junior Bairro, 2012.

2ª Stage

Cut out the paper where indicated (   ) and fold (   ) slightly the parts of the whole cube. Figure 2 shows the structure of the paper after cutting and folding.

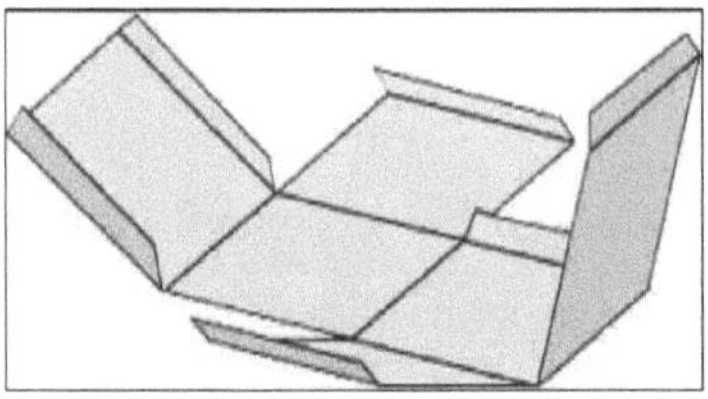

Figura 2- Representational model of the cubic figure after the cuts

Source: Junior Bairro, 2012.

3 ª Stage

Write the desired information on each side of the cube, for example: symbol of the chemical element, nomenclature, atomic number, physical state, employability, electronic distribution, among others.

4ª Stage

After the information has been inscribed on the six sides of the cube, the inside and outside of the cut-out are covered with contact paper to increase durability and prevent water absorption (humidity).

5ª Stage

Fold and mould the cubes to fit them into the compartments of the support (wooden box).

2.1.5 Lesson development and data collection instrument

When it comes to eighth grade science content, the Basic Education Curriculum Guidelines (2008) recommend an introduction to the study of chemistry, focusing on matter: structure and classification, in a contextualised and interdisciplinary way, covering topics such as matter and its transformations (physical and chemical), the constitution of matter, an introduction to atomic theories, the Periodic Table, the formation of matter from the union of atoms (chemical bonds), the study of substances (acids, bases, salts and oxides), (SANTANA, 2009).

When dealing with the topic of the Periodic Table and Chemical Bonds, theoretical lessons were planned, proposing an approach to the specific content in a contextualised way with everyday life, which favoured the study of the constitution of the bodies that present themselves to our perception as well as their constitution, going beyond what the student initially sees, feels or touches. The sequence of the approach was: Periodic Table, History, Chemical Elements and Organisation, some periodic properties (atomic radius, electronegativity, electropositivity) electronic distribution, and the way elements come together to form a substance (Chemical Bonds).

This work was carried out at the Professor Anacleto Damiani Basic Education School with 37 Correção de Fluxo students, 16 of whom were in class A and 21 in class B.

Group A took part in the lessons and in solving the proposed activities and used the traditional periodic table.

The students in group B carried out their activities using the alternative periodic table made by the teacher in an interactive way in groups at to solve examples and exercises. The way the table was used was as follows: the Alternative Periodic Table was placed in the middle of the classroom and used by groups of two to three students.

The exercises relating to the classification of elements into groups, periods and chemical bonds were solved by the two groups using the electronic distributions of the chemical elements proposed in the exercises, later checking the results against the data in the traditional Periodic Table (class A) and in the Alternative Table (class B).

Based on the observation of the electronic configuration of the valence layer of each atom, they proposed the Chemical Bonds required for certain compounds such as sodium chloride (table salt) and hydrochloric acid (acid present in the stomach that helps digest food), among other compounds used or present in the student's daily life, classifying the bonds of these substances as ionic or covalent.

While solving the proposed activities, each group solved the questionnaire below:

2.1.6 Questionnaire applied:

1. How old are you?
2. Do you like studying chemistry?
3. Do you have an easy time understanding chemistry?
4. Does chemistry play an important role in everyday life?
5. Using the periodic table, indicate the group and period of each element:

Elements	Group	Period
K		
F		
N		
AI		
Ba		
S		

6 - Based on the electronic distributions, indicate the group and period of each element:

Elements	Electronic Distribution	Group	Period
In			
Ca			
P			
0			
Cl			
Ne			

7- Propose the Lewis structures for the compounds below and classify the bonds as ionic or covalent.

NaOH	LiOH	HC1	HF
KCL	CaO	BaS	

Labour data, 2012

After solving the exercises, fill in the table below according to the categories proposed by the student.

I don't know ->**You had no idea how to solve the** proposed exercises;

I was in doubt-" **You've mastered the subject relatively well, but you were in** doubt between one way or another and ended up doing it the wrong way;

I didn't understand -" **you know the subject, but you didn't understand how to solve the question;**

You know how to do it, but due to nervousness, anxiety, fear or any other emotional factor, you haven't managed to solve the problem;

There wasn't enough time -" **you would have known how to solve** it, **but there wasn't enough time.**

I know - "**I got it right, because it was easy.**

QUESTION 5 - Using the periodic table, indicate the group and period of each element.

	K	F	N	AI	Ba	S
I don't know						
I stayed at Doubt						

White						
I don't understand						
There was no time						
I know						
Group						
Period						

QUESTION 6 - Based on the electronic distributions, indicate the group and period of each element.

	In	Ca	P	0	Cl	Ne
I don't know						
I was in doubt						
White						
I don't understand						
There was no time						
I know						
Group						
Period						

QUESTION 7 - Propose the Lewis structures for the compounds below and classify the bonds as ionic or covalent.

	NoH	KC1	HC1	HF	KC1	CaO	BaS
I don't know							
I was in doubt							
White							
I don't understand							
There was no time							
I know							
Connection type							

2.2 LITERATURE REVIEWS

2.2.1 Science teaching - legal guidelines

Concern about education and the methods used to teach chemistry in primary school (eighth grade) can be seen as a basis for developing chemical thinking, as well as chemical language and writing. Traditional teaching methods, based on lectures, still lead the way in the classroom and the excess administered by institutions means that science teaching becomes content-based (TASCA, 2006).

> According to the PCNs, the teaching of the structure of matter should be covered in the 4th cycle of primary school, where themes are developed that can integrate this topic. The proposed content is divided into 4 thematic axes: **Earth** and **Universe**, **Life** and **Environment**, **Human Beings** and **Health**, **Technology** and **Society.** These themes make up the rhythm of primary education and studies on the structure of matter should cover the basic concepts that involve chemistry, working with thematic proposals that relate to

> environmental, health and social problems and various other themes that can be cross-cutting. In this cycle, science studies the processes linked to the earth's composition and the cycles of materials and energy flow, whether they occur naturally or as a result of human action. In the case of the approach to the structure of matter, it is not necessary to delve into the molecular or atomic level, but it should be understood that matter is made up of particles, such as atoms or molecules. In addition, an elementary study of chemical equations can begin, considering aspects such as: what the reactants are, the reaction conditions and their products, which is already complicated for this level of schooling (PCNS, accessed 03/05/ 2013).

2.2.2 Elements of pedagogical practice for teaching science

According to Cachapuz (2000), science education must ensure that the learning of knowledge becomes useful and usable in the student's daily life, and there is an increasing call for inter- and transdisciplinarity, the approach to everyday problem situations that allow knowledge to be built up and reflection on science processes, the interrelationship between science/technology/society/environment, methodological pluralism and formative rather than classificatory assessment.

2.2.3. The problematising approach

Progressive pedagogy uses the problematising teaching methodology as the main method for critical reflection on reality, promoting knowledge.

"Concepts such as problematisation, dialogue, freedom and conscientisation are key elements in Paulo Freire's progressive teaching" (MIRANDA & BARROSO, 2004, p.634).

> According to Freire (2004, p.47), teaching is not the simple transmission of knowledge about the object or content. This is done much more through the pure description of the knowledge of the object to be mechanically memorised by the students. Teaching to learn is only valid from this point of view, it should be repeated, when students learn to learn by learning the raison d'être of the object or content. For this reason, teaching is an act that involves creation and critical thinking about the various aspects of reality. The curiosity of teacher and student, in action, lies at the basis of teaching and learning.

According to Brasil (2003e, p.45), the path of problematisation in educational practice, as we believe, is based on dialogue and seeks to break the vertical way of teaching, "no longer anaesthetising students, but stimulating their creative power and understanding of **reality in permanent transformation".**

According to Berbel (1998, p.142), problematisation is a teaching, study and work methodology that can be used whenever appropriate, in situations where the themes are related to life in society.

There seems to be a consensus among teachers of the natural sciences and chemistry that the use of experimental activities should permeate teaching-learning relationships, since they stimulate student interest in the classroom and engagement in subsequent activities (GIORDAN, 1999; LABURÚ, 2006).

As Carrascosa (2006) argues, experimental activity is one of the key aspects of the science teaching-learning process. Therefore, to the extent that experiments are planned with which it becomes possible to strengthen the link between motivation and learning, it is hoped that student involvement will be more vivid and lead to positive results in the evolution of knowledge.

2.2.4 The interdisciplinary relationship

The use of this approach is provided for in the teaching reference document of the Ministry of Education and Culture - MEC - (BRASIL, 2000) when it points to interdisciplinary work as "cultural sharing", capable of promoting relations between Science and Technology in the school environment, favouring an understanding of the history of the sciences as a whole, giving context to reality and transcending the domain of discipline.

According to studies carried out on the subject of interdisciplinarity in the educational context (FAZENDA, 1992; SANTOMÉ, 1998), there are variations in conceptual understanding and the absence of a model that systematises its practice. Thus, due to the absence of an understanding of its historical significance and the lack of methodological systematisation, multiple conceptions and possibilities for implementation are found today.

The interdisciplinary theory approached by Fourez (1995) is based on "specific practice aimed at addressing problems relating to everyday existence", with the aim of promoting Scientific and Technological Literacy in education. Thus, as an interdisciplinary model, the Island of Rationality is aimed at the political practice of the student, through the negotiations and decisions made during its development, with the aim of "disseminating sufficient knowledge to the population so that the decisions of technicians can be sufficiently understood and also democratically controlled. It's about giving responsibility to society".

When teachers are able to bring in knowledge from other areas when working on specific knowledge in their discipline, they begin to blur the boundaries between disciplines. For this to happen, the teacher needs to have a minimum knowledge of the other disciplines with which they will establish the appropriate relationships. In this way, as RIOS (1997) argues, this practice can be considered an attempt to broaden knowledge, and it is only possible to talk about interdisciplinarity if there is, in fact, a creative dialogue between two or more disciplines.

> In this sense, interdisciplinary pedagogical practice, as discussed by FAZENDA (1979), must involve the exercise of relationships of association, collaboration, co-operation, complementation and integration between disciplines. According to this author, exercising interdisciplinarity means renewing and restructuring pedagogical work through integration between disciplines.

2.2.5 The research

The research method consists of getting students to observe, collect data and reflect on it. This data can come from spontaneous manifestations or provoked, from nature, society or man.

Research, in a way present in all the methods presented, is the
basic point of what will be explained below (NERICI, 1989. p.148).
With three types;

- ***P*** Bibliographical research: emphasises research in books, which provides results to explain and understand the subject.
- ***P*** Field research; emphasises surveys and questionnaires, which go straight to the source, through testimony, in order to better notice, feel and observe it.
- ***P*** Experimental research; this is research that collects data with the aim of proving a previously developed hypothesis.

Research is the method by which students learn to discover new worlds and awaken their curiosity. In chemistry, this is very important because it is from there that the best discoverers and scientists emerge.

2.2.6 Scientific reading

A scientific reading is a reading aimed at analysing the work.

Unlike pleasure reading, scientific reading looks for the following aspects:

Central idea

v' Subsequent ideas

Arguments

F They strengthen and confirm the central idea.

It is very important in scientific reading to know who the author is in order to easily extract the central idea.

Before writing a summary or review, it's important to learn how to do scientific reading.

According to Silva (1998), the science teacher is also a reading teacher. In other words, it can be assumed that they are also responsible for providing opportunities for students to practice reading in the classroom. This is because all school subjects are based on written language. Ricon and Almeida (1991) point out that this practice makes it possible to connect students' lives with science, so that this relationship is intensified in pedagogical interaction.

2.2.7 Group activity

> The group study method consists of two or more students studying a topic, a unit or carrying out a task together. Group study is increasingly being applied at all levels of education, as it has proved to be an excellent way of carrying out instruction and social training (NERICI, p.194. 1989).

group activities allow:

D Developing sociability;

D Developing the ability to liaise and work with other people;

F Favour learning;

F Strengthen group spirit;

The group study method can be developed in two ways.

T All groups studying the same topic or carrying out the same task;

C Each group studying part of a topic or carrying out part of a task, or even studying different topics and carrying out different tasks;

It's important, because they help each other, clear up doubts and help each other, where socialising and studying are interesting aspects of our students' lives.

2.2.8 The observation technique

Observation is an important and fundamental factor in an individual's growth, due to the awareness of external and internal reality that it can provide. All knowledge and all human action, in order to be adequate, depend fundamentally on observation, which should be the starting point for reflecting on reality and acting on it so that the results are more likely to be correct.

Hence the need for educational action at home and at school to develop the individual's capacity for observation.

Observation is the act of the spirit of apprehending (perceiving) an external or internal phenomenon and recording it as objectivity, with which we develop behaviours of curiosity, reflection and research (NERICI, p.353. 1989).

The use of experimental activities is a teaching practice that shows the relationship between theory and experimental results, which makes it very productive as it provides students with models for observation, reasoning and interpretation. Through this teaching strategy it is possible for students to form their own scientific criteria, where they use their theoretical knowledge and intuition to arrive at an understanding of the experiments, in other words, reinforcing learning.

2.2.9 The experimental activity

It consists of bringing students into contact with a phenomenon that motivates them and leads them to reproduce it, improve it, get to know it, master it and use it.

> Experience represents man's purposeful action on physical, social and human nature. Children act experimentally to see what happens. Adolescents do it with the intention of increasing their personal power, as a form of affirmation. Adults also experiment, but more to fulfil a need that preoccupies them. Experimentation, didactically, is the extraordinary

technique for active and motivating teaching, which requires the full participation of the learner (NERICI, p.170. 1989).

Practical lessons are an efficient way of teaching and improving the understanding of chemistry content, facilitating learning. Experiments make it easier to understand the nature of science and its concepts, help develop scientific attitudes and diagnose non-scientific conceptions. They also help to arouse interest in science. With this in mind, we seek to motivate students to take part in practical lessons, making them more dynamic and enabling them to develop the content according to their wishes and in a more meaningful way (ALMEIDA, et al, 2013).

2.2.10 Instructional resources

It is the set of learning activities planned with a view to facilitating the student's learning, according to their real possibilities so that they can achieve the objectives set, including, for this purpose, learning alternatives, with a view to better adapting the studies, aimed at efficiently and adequately guiding the student so that they can achieve the objectives and teaching in a satisfactory way, for example, standing with your back to the student for as short a time as possible, and in order not to damage the presenter's vocal cords, you should avoid talking while writing so as not to ingest the chalk dust. The magnetic board (whiteboard) avoids this problem and can also be used to affix objects or papers using a magnet. The flip chart is a resource made up of an easel and blank sheets of paper and is used to take momentary notes during the presentation of the subject. It differs from the serial album, which consists of a sequence of properly organised sheets containing information to help understand the content being explained. Transparency, on the other hand, requires special equipment (overhead projector) to project the material onto a screen. The use of films and slides makes the presentation closer to reality, as it allows existing situations to be shown, the former being more dynamic than the latter, allowing contact with the characteristic movements and sounds that occur in the situation under study, which cannot be achieved through the slide (CABRAL and MIRANDA,2001,p.4).

2.2.1.1 THE STUDY OF THE PERIODIC TABLE

2.2.1.2 The evolution of the Periodic Table

The chemical elements are arranged in a table in proportion to their characteristics and properties.

For example, it allows us to predict the behaviour of atoms and the molecules they form, or to understand how certain atoms are extremely reactive while others are practically inert, etc. This is thanks to the discovery of individual elements.

The first discovery of an element took place in 1669, discovered by Henning Brand, the element being phosphorus.

The first classification was the division of elements into metals and non-metals.

This made it possible to anticipate the properties of other elements, thus determining whether or not they were metallic.

The first list of chemical elements was released by Antoine Lavoisier in 1789 and contained 33 elements. Although Lavoisier grouped the elements into gases, metals, non-metals and earths, chemists spent the next century searching for a more precise construction scheme.

The Periodic Table is an essential part of the language of chemistry. It has much in common with a treasure trove, offering a guide to the similarities and differences between elements. Because of the way the elements are arranged in the Periodic Table, we can predict their behaviour and write chemical formulas for compounds using just a few general guidelines. Using these rules is not the same as understanding why elements in certain areas of the Periodic Table behave as they do, but the trends that emerge from the arrangement of the elements in the Periodic Table allows a chemist to recall useful facts about the types of compounds formed from specific elements and their chemical reactions. It also organises the elements in a logical way, presenting them in families and groups that have similar characteristics that allow the relationship between different elements to be compared more easily.

2.2.1.3 Johann Dobereiner's triads

In 1829, Johann W Dobereiner published the Triads, in which the elements were separated into groups of three, with similar properties and in such a way that the central element of the Triad had its atomic mass approximately equal to the arithmetic mean of the masses of the other two. Some existing metals were unable to adapt to the triads (GONÇALVES, 2003, p.80).

2.2.1.4 Leopold Gmelin's system

> SANTOS and MOREIRA, (2011) The German chemist Leopold Gmelin worked with this system and by 1843 he had identified ten triads, three groups of four, and one group of five. Jean Baptiste Dumas published a paper in 1857 describing the relationships between the various groups of metals. Although several chemists were able to identify relationships between small groups of elements, there was still no scheme capable of covering all of them.

The German chemist August Kekulé observed in 1858 that carbon has a tendency to bond with other elements in a ratio of one to four. Methane, for example, has one carbon atom and four hydrogen atoms. This concept became known as valence.

2.2.1.5 The Telluric Bolt by Alexandre Emile Béguyer de Chancourtois

In 1862, the French geologist, in an attempt to compile a table of chemical elements, produced a work in which the elements were arranged around a cylinder (spiral), divided into 16 parts.

Chancourtois Cylinder

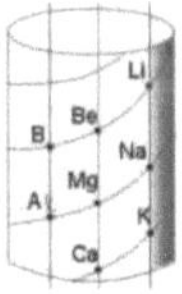

Source: LANAO (2005)

The elements were distributed in order of their atomic weights, calculated on the basis of the Oxygen 16 standard. With this distribution, elements such as oxygen, sulphur, selenium and tellurium remained on the same vertical line. Despite the success of this group, other groups brought together elements with different properties such as boron, aluminium, nickel, arsenic, lanthanum and palladium. Chancourtois discovered periodicity at the wrong frequency (GONÇALVES, 2003, p.81).

2.2.1.6 William Odling's Table

William Odling, British chemist, died aged 92. His research took mass and properties into account.

In his table, Odling arranged the elements in ascending order of their atomic weight , leaving empty spaces for elements that might be discovered. Odling even solved the problem of tellurium (Te) and correctly placed mercury (Hg) and platinum (Pt), discoveries that Mendeleyev failed to make in his first attempt at drawing up the table. the Periodic Periodic Table.

However, Odling received little recognition for his work, as there were suspicions that he, as secretary of the Chemical Society of London, had contributed to discrediting the chemist John Reina Newlands, who wanted to see his periodic table published.

2.2.1.7 The John Alexander Reina Newlands Table

An English chemistry professor, in 1864 he suggested that the elements should be placed in a periodic model in eight groups (similar to musical notes), in ascending order of their atomic masses. The group of sodium, lithium and potassium was well set up, but elements such as chlorine, bromine, iodine, iron and copper were left out of his work. Atomic numbers and quantum theory were unknown until 1864. Protons were only discovered in 1886 (GONÇALVES, 2003, p.81).

2.2.1.8 The Julius Lothar Meyer Table

Meyer's table included approximately half of the known chemical elements, the distribution was made as before, in ascending order of their atomic weights, and demonstrated that chemical and physical properties were periodic functions. Empty spaces existed in the table, and in 1868 he made changes to his work and formulated the extended Periodic Table. He is only not considered the father of the Periodic Table because he delayed publishing it (GONÇALVES, 2001, p.27).

2.2.1.9 Dmitri Ivanovich Mendeleev's Table

According to GONÇALVES, (2001, p.28) the so-called creation of the Periodic Table had its crucial moment on 17 February 1869, when Mendeleev, studying , used the increasing order of their atomic weights, so that elements with similar properties would remain in the same group. After that, Mendeleev built cards, each one representing a chemical element, where he listed the known properties and put them in ascending order of atomic weights to try to find out how many vertical groups he needed. The only table with a vertical, horizontal and diagonal relationship.

2.2.1.10 Julius Lothar Meyer X Dmitri Ivanovich Mendeleev

> SANTOS and MOREIRA (2011) Russian chemistry professor Dmitri Ivanovich Mendeleiev and Julius Lothar Meyer independently published their periodic tables in 1869 and 1870 respectively. Both constructed their tables in a similar way: listing the elements in a row or column in order of atomic weight and starting a new row or column when the elements' characteristics began to repeat.

2.2.1.11 William Ramsey

> He suggested that Argonium be placed between Chlorine and Potassium, forming a family with Helium, but this ended up going against the distribution of the elements in ascending order of their atomic weight. This gave rise to a new periodic group, called the zero group, because the valence of the elements in this group was equal to zero (GONÇALVES, 2001, p.86).

2.2.1.12 The Table after Henry Gwyn Jeffeys Moseley

He determined the wavelengths of alpha radiation from various chemical elements, obtaining the relationship between the wavelength emitted and the number of protons in the elements that emit it. And in 1913, the so-called Atomic Number nuclear charge emerged and was used to order the elements in the Periodic Table.

2.2.1.13 The Table after Gleen Theodore Seaborg

US physicist Glenn Seaborg led the research team that discovered plutonium in 1940, and in 1941 isolated uranium-233.

He oversaw plutonium manufacturing and enrichment research for the Manhattan Project, which culminated in the development of atomic weapons.

Seaborg and his colleagues also discovered americium, berkelium, californium, curium, einsthenium, fermium, mendelium, and nobelium, and identified more than 100 isotopes of elements along the Periodic Table.

He shared the 1951 Nobel Prize for Chemistry with Edwin M. McMillan, and he is the namesake of Seaborgio, discovered in 1974 by Albert Ghiorso.

In 1963 he acted as a key scientific and diplomatic negotiator in the work that led to the Nuclear Test Ban Treaty (1963), which limited the testing of nuclear weapons in the atmosphere and under the sea, and the Nuclear Non-Proliferation Treaty (1968), which limits the spread of nuclear weapons technology.

2.2.1.14 The Periodic Table according to IUPAC

According to Mendes (2011), although the Mendeleev/Meyer table demonstrated the periodic nature of the elements, it was only in the 20th century that explanations were found for why the properties of the elements varied periodically. The development of atomic models and quantum theory made it possible to rationalise knowledge of the properties of the elements and arrive at the configuration of the current periodic table. This has been expanded over time as new elements have been discovered. The current table contains 118 elements, arranged in horizontal lines (periods) and vertical lines (groups), in ascending order of atomic number. The horizontal rows are arranged so that elements with similar properties are in the same columns (groups or families). The group is considered the most important method of classifying elements. In some groups, elements have very similar properties and exhibit a clear trend in properties across the group. These groups have been given trivial names, for example, the alkali metals, alkaline earth metals, halogens, noble gases, and so on. Some other groups in the periodic table show a lesser degree of similarities/vertical trends and are referred to simply by their group number. Although groups are the most common way of classifying elements, there are areas of the periodic table where horizontal trends and similarities in properties are more significant than vertical trends. In the Periodic Table, each element is shown with its symbol and atomic number. Many versions of the table also show other atomic properties and physical properties. Throughout the 20th century, alternative representations of the Periodic Table appeared, mainly for didactic reasons. However, the **"traditional" Periodic** Table **remains** the representation of the mathematical distribution of chemical elements according to their properties.

The Periodic Table is now ubiquitous, providing a useful framework for classifying, systematising and comparing the many different forms of chemical behaviour. The Table has found many applications in chemistry, physics, biology, engineering and materials science.

2.2.1.1.1 ALTERNATIVE METHODOLOGIES FOR STUDYING THE PERIODIC TABLE

"Studying the Periodic Table is always a challenge, as students find it difficult to understand the periodic and aperiodic properties and even how the elements were arranged in the table and how these properties relate to the formation of substances. In most cases, they don't know how to use it and end up thinking that the best way is to memorise the most important information. In the search for teaching materials to teach periodic and aperiodic properties, **it can be seen that the most commonly used methods are traditional (textbooks)".** (GODOI, **et al**, 2009).

There is no doubt that the teacher's didactic experience and creativity can make teaching chemistry more enjoyable and with good results in terms of student learning (MAGALHÃES, 2002). The difficulty students have in forming mental models to understand chemical concepts and phenomena can be overcome through games, illustrations and pictures, and the use of computers in the classroom certainly makes it easier to understand chemical concepts (FERNANDES, 2007).

In this context, we can cite some works related to the study of the periodic table that show positive results:

a) The work carried out by **Ribeiro (2010) was centred on the Universe".**

Once the theme had been defined, about three months in advance, they had free time each week to discuss and organise the event. During this period, as well as guidance, the students researched websites that presented experiments and curiosities in the area of chemistry and prepared notebooks with experiments selected by the teacher in charge. The Cultural Fair was held on 11 December 2009. The day before, the laboratory was organised so that the experiments could be set up on the side benches, the posters showing curiosities were fixed to the wall and the Chemistry game was set up. Seven experiments were set up and fifteen students were assigned to them. Two students were responsible for the trivia and one student and the teacher were responsible for the game.

The conclusion was satisfactory because it aroused the students' curiosity with something practical and not too time-consuming. It was also possible to observe that there were few students who did not participate effectively in the activities, mobilising the school teachers and management.

b) Playful activities.

According to DANTAS (2010), her work makes it very clear why we should work with playful activities. She gives us examples of different ways of working that require a lot of attention, motivation and interest. Among the various methodologies she presents, we can mention a few: - Mastering organic chemistry;

- Nomenclature of alkanes;
- Thermochemical ludo;
- Soaking water;

Among other things, I'm arguing that our students need different lessons to attract their attention, to motivate them more, so that chemistry doesn't become a seven-headed beast.

c) Educational games.

According to the PCNs (BRASIL, 1998), chemistry as a school subject is an instrument of human formation, a means of interpreting the world and interacting with reality. Understanding chemistry content is linked to a new vision of science and scientific knowledge.

It is therefore believed that educational games can be a new and important resource for teachers and can show problem-solving skills that favour learning and adolescence.

Games can be a way of sparking student interest, motivation and improving learning.

2.3 ANALYSES AND DISCUSSION OF RESULTS

The main aim of this work was to see if using an alternative methodology in science and chemistry teaching with a representational model of the alternative table would contribute significantly to the mediation of teaching by the teacher and the appropriation of knowledge by the student. The results obtained are shown below.

2.3.1 Production of recyclable paper and making the Periodic Table Alternative

1ª Stage - allocation of cans for paper collection

Two bins for the separate collection of plastic and paper were placed next to the door of each classroom, as shown in figure 1.

Figure 1: Selective collection of paper and plastic

Source: Labour data, 2012.

2ª Stage - Paper preparation

After collecting and separating the paper, it was cut into small pieces for subsequent shredding (figure 2).

Figure 2: Paper preparation
Source: Labour data, 2012.

3ª Stage - Paper shredding

After sorting and cutting, the paper was left to soak for 24 hours and then shredded using a blender (figure 4).

Figures 4 - Paper shredding

Source: Labour data, 2012.

4ª Step - Removal of excess water

A conventional kitchen sieve was used to remove the excess water. This procedure is necessary so that the paper doesn't become full of holes and uneven (figure 6).

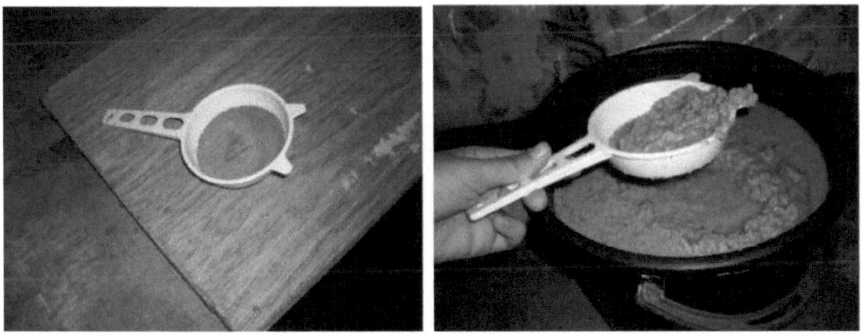

Figures 6 - Removing excess water
Source: Labour data, 2012.

5ª Step - Preparing the sheets of paper

To prepare the sheets, a strong wooden base had to be used so that the sheets were straight and even. This base was covered with denim fabric to accommodate the dough for forming the sheet. This procedure was repeated six or seven times, placing one layer on top of the other (jeans/paper dough), and another wooden base was placed on top of the last layer to form a press (figure 8).

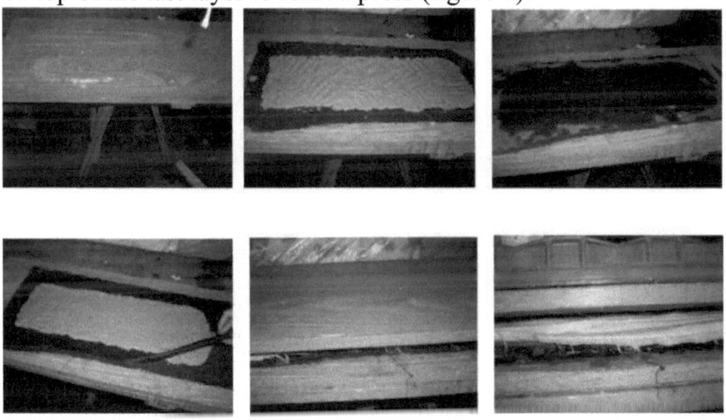

Figure 8 - Preparing the sheets of paper

Source. Labour data, 2012.

6ª Stage - Drying the paper

After the paper had been in the press for 24 hours, the sheets still stuck to the jeans were carefully removed to avoid breakage. The sheets were dried in the shade (figure 9), because when dried in the sun as indicated in the method, they end up deformed and become very brittle.

Figures 9 - Paper drying

Source: Labour data, 2012.

7ª Stage - Paper moulding

After 48 hours of drying, the sheets naturally loosened from the jeans and were then put through the electric sieve three times until they were very thin, uniform and compact (figure 10).

Figure 10 - Moulding the sheets of paper

Source: Labour data, 2012.

8ª Stage - Assembling the cubes to make the periodic table

To build the 118 cubes, which represent the chemical elements of the periodic table, it was necessary to work with one sheet at a time: contact paper was applied to one side of the sheet of recyclable paper; with the help of the mould, several cubes were drawn on one sheet; the information about each chemical element was written starting with Hydrogen(1) and ending with ununoctium (118) on each side; contact paper was applied again; and the cut-outs were made; the cube was assembled by placing them in a wooden box, made especially for the Alternative Periodic Table, figure 11.

Figure 11 - Alternative table

Source: Labour data, 2012.

2.3.2 Comparative results obtained using the Traditional Periodic Table and the Alternative Table

Graph 01 shows the total number of remedial students at the Professor Anacleto Damiani Primary School involved in the study of the Periodic Table:

T Class A - 16 students solved the exercises using the traditional Periodic Table.

T Class B - 21 students who carried out their activities using the alternative Periodic Table made by the teacher.

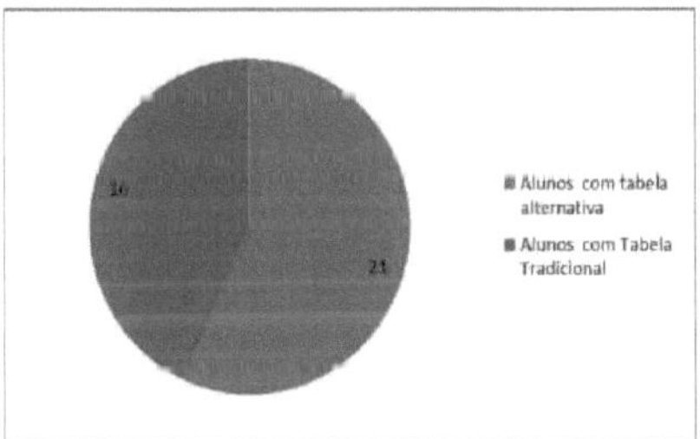

Graph 01: Total number of participating students

Source: Labour data, 2012.

Graphs 02 and 03 show the age range of the Flow Correction students.

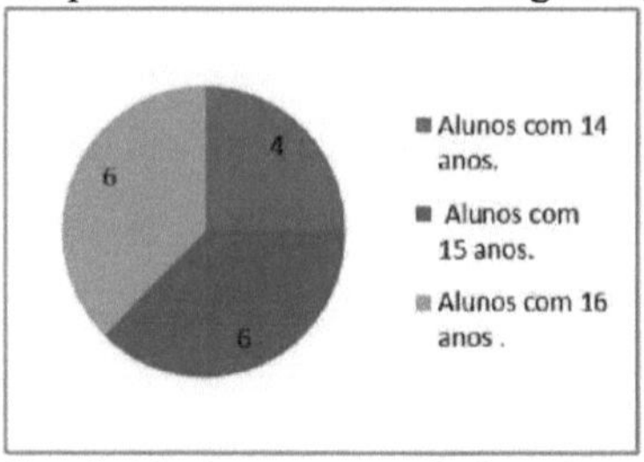

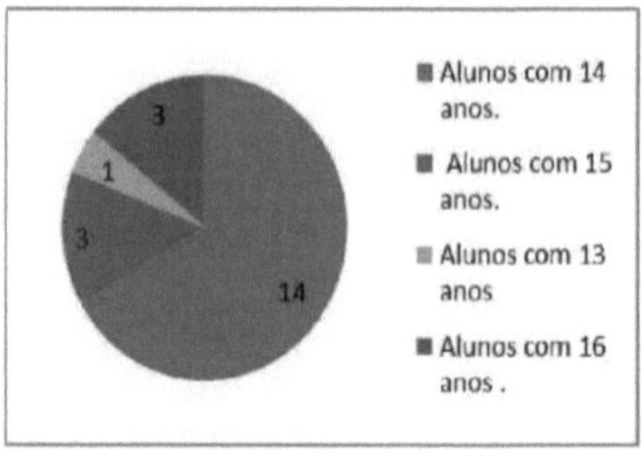

Graph 02: Age group of Flow Correction students in class A.
Graph 03: age group of students in class B

Source: Labour data, 2012.

As this is a special Flow Correction class (students behind in terms of grade/age), most of the students are not in the ideal age group for the eighth grade of primary school (12 to 14 years old). The majority of the students in class A (12 students) should already be in secondary school, but because of the failures they are behind in terms of grade/age. Some students aged 16, for example, are attending the eighth normal grade or the Correção de Fluxo grade for the second and third time.

In class B it can be seen that 15 students are in the correct age group in relation to the grade they are studying, 13 to 14 years old. The 03 students aged 15 were repeating the grade for the second time and the 03 students aged 16 were repeating the grade for the third year.

In this sense, it is up to the competent bodies to stimulate debate and reflection on the actions developed by the players involved in correcting the gap, as well as the specific assessment process for correcting the age/grade gap of students who are lagging behind (LDB 9.394/96, Art. 24, item V, b).

Graphs 04 and 05 show the results of **the question "do you like** or **dislike studying chemistry"**, obtained during the work carried out with grades 8ª A and B of Flow Correction.

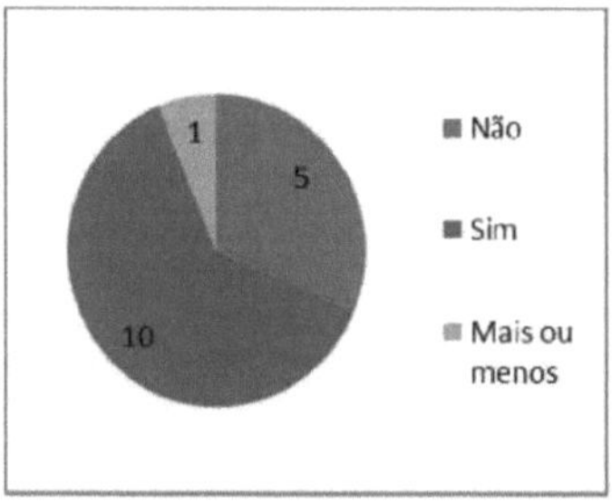

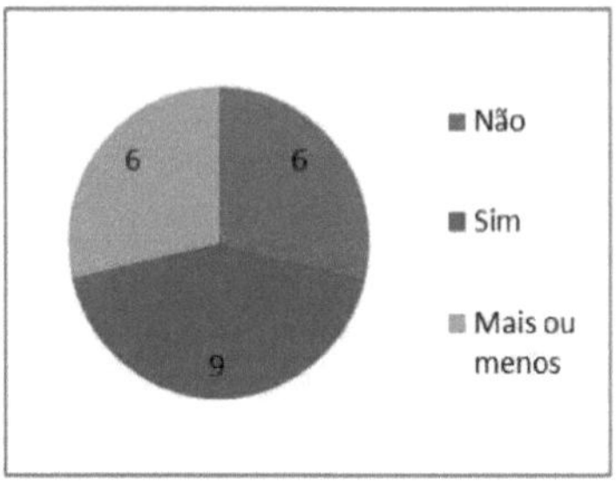

Graph 04: Do you like studying chemistry? Class A.
Graph 05: Do you like studying chemistry? Class B.
Source: Labour data, 2012.

In class A, 10 students confirmed that they liked studying chemistry, as follows: "[...] Well, I like chemistry, but I don't know why I find it so difficult to learn, even though I practise [...]". This picture makes us believe in the importance of the teacher's perception in the classroom, leading him or her to analyse why these students can't learn, even if they want to. We know that there are many factors involved in good science learning, but similar studies have shown that one of the most frequently mentioned factors is the methodology used by teachers (DUARTE **et al**, 2010). These 10 students still regret that they have little time, little content, few experimental lessons and that most of the time the content is not contextualised with everyday life. The other students said they didn't like chemistry and only 1 said he liked it more or less.

In class B, of the 21 students, 09 said they liked studying chemistry. 06 students said they liked chemistry more or less and 06 said they didn't like chemistry, giving reasons and asking questions such as: [ ] The content is very difficult, abstract.

[..] I can't understand. [...;] Where will I use it? They said that they are
These results are in line with a study carried out by CARDOSO, (1998), who in "an analysis of 157 students came to the following conclusion: the justifications given by the 25% who said they did not like studying chemistry. These results are in line with a study carried out by CARDOSO, (1998), **who "analysed 157 students** and came to the following conclusion: the justifications given by the 25% who said they didn't like studying chemistry were classified into four groups, according to the author:

1- The first group (53 per cent of the responses) presents the students' view of the subject and the way it is taught. For them, chemistry has too many subjects to be studied and memorised, as well as topics that are considered abstract or taught in a confusing and superficial way.

2- - The answers from the 2nd group (17%) reveal a lack of interest on the part of the students, who think they already know enough, turning to specialised magazines when necessary in order to obtain more information. Some students claim that they won't need chemistry in their future profession, and that they don't find these concepts useful.

3- In the 3rd group (22 per cent of responses), the lack of motivation to study is associated with the difficulty in understanding and assimilating chemical concepts.

4- In the 4th group (8% of responses), dislike is linked to the subject studied or the presence of concepts from disciplines such as maths and physics, although chemical knowledge is considered important and useful

Graphs 06 and 07 show **the results of the question "Do you find it easy to learn chemistry?", obtained during the work carried out with** ***the*** **8ª** grade A and B of Correction de Fluxo.

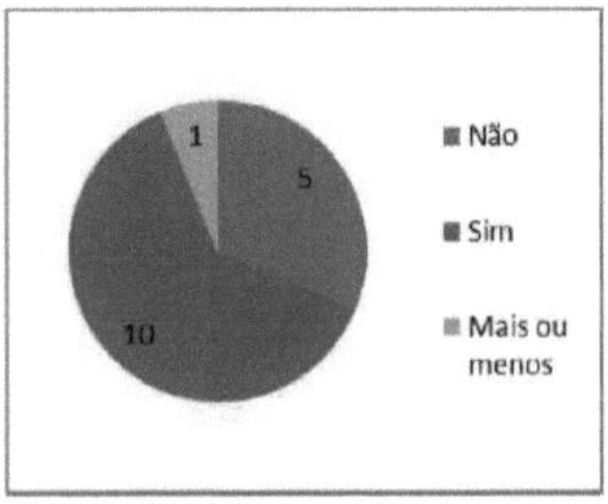

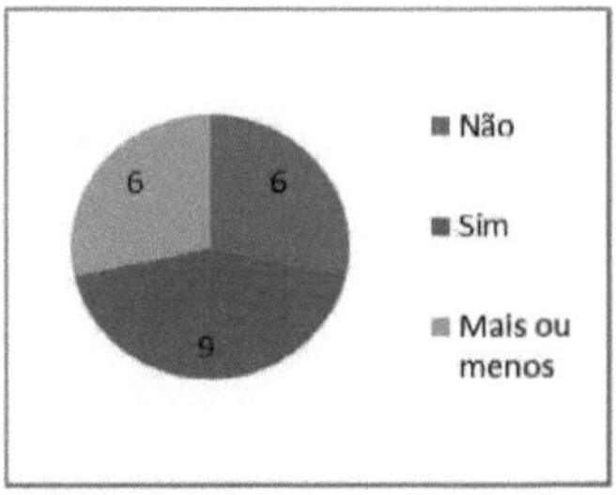

Graph 06: Do you find chemistry easy to learn? Class A.

Graph 07: Do you find chemistry easy to learn? Class B.

Source: Labour data, 2012.

Of the 16 students in class A, 10 say they find it easy to learn chemistry, about 62.5%, but on the other hand 31.25% or (5 of the 16 find it difficult to teach chemistry). In class B, of the 21 students, 9 say they find it easy to learn chemistry, about 42.85%, but 28.57% find it difficult (6 of the 21 students). From a brief analysis, we can see that the students who have more affinity with maths are the students who do best in chemistry, they do well because in observations made during the work [...] they confirm that they awaken curiosity for differentiated work and interest in knowing more about the content when it involves logical reasoning.

Graphs 07 and 08 show the **results of the question "Chemistry plays an important role in everyday life", obtained during *the* work carried out with** grades **8ª** A and B of Flow Correction.

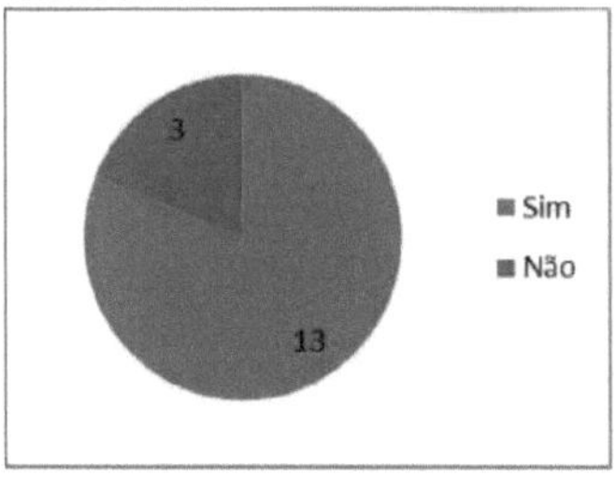

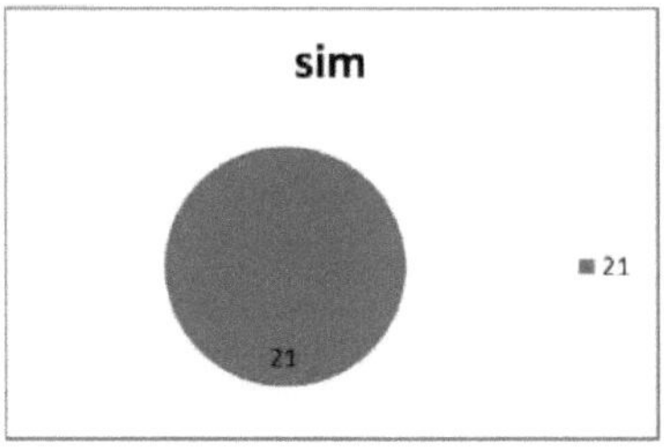

Graph 07: Does chemistry play an important role in everyday life? Class A.

Graph 08: Does chemistry play an important role in everyday life? Class B.

Source: Labour data, 2012.

It can be seen that the majority of students in both classes believe that chemistry does play a fundamental role in their lives. Because it's present in medicines, clothes and shoes. [...] they say that chemistry is fascinating, and trying to understand chemical processes, how transformations occur, is not always easy to understand, but it awakens the imagination, something beyond the hands.

A study carried out by CARDOSO, 1998, with basic education students (eighth grade and high school) shows that:

a) **"Chemistry was considered important in the personal lives of 76 per cent of the students,** who showed that they were well aware of its presence in everyday life. In the examples cited, chemistry is related to metabolism (respiration and digestion), to nature (in forests, fire, air and water), to substances used in everyday life (medicines, cosmetics, cleaning and hygiene products, food, plastics, fuels, fabrics and paints) or to processes that help and improve human living conditions (combustion in cars, industrialised food, recycling waste and paper).

b) The percentage of 3rd graders who answered yes to this question (84 per cent) was higher than those in the 8th grade (63 per cent), suggesting that the years of study have broadened their knowledge and perception of the presence of chemistry **in their lives. "It was observed** that the students did not have a more in-depth discussion about the role of chemistry in the examples cited, raising the hypothesis that an effective relationship between school chemistry and **everyday chemistry** was not established."

This situation can and should be changed by the way the teacher approaches this content, content that can and should be contextualised with everyday life without losing its scientific character, remembering that these students are only in the eighth grade and that this class will have a long way to go during secondary school.

The following results are related to the activities carried out during Science (Chemistry) lessons. After solving the proposed exercises, the students indicated the options that best suited the difficulty of solving them, the options being: I don't know, I was in doubt, I didn't understand, I blanked, I didn't have time or I already know.

Graphs 09 and 10 show **the results of the question** "Difficulty students had in classifying different chemical elements into groups and periods", **obtained during the work carried out with** grades **8ª** A and B of Flow Correction.

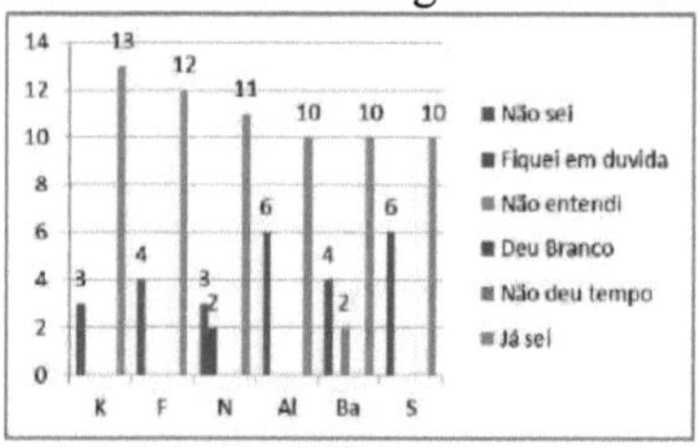

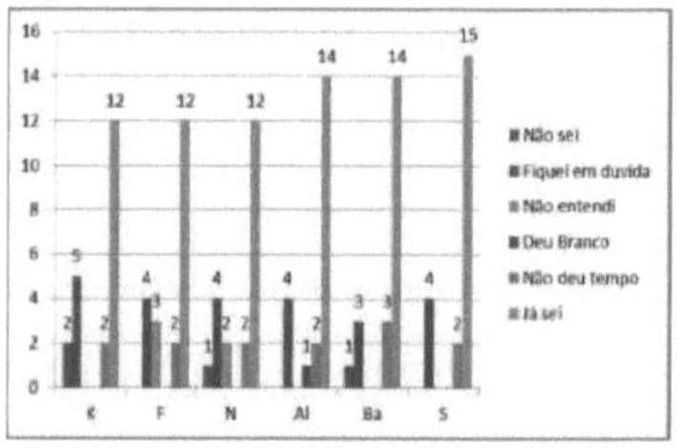

Graph 09: Students' difficulties in classifying different chemical elements into groups and periods - Class A.

Graph 10: Students' difficulties in classifying different chemical elements into groups and periods - Class B.

Source: Labour data, 2012.

In class A, 68.75 per cent (an average of 11 students) said they already knew how to identify the group and period of a chemical element and 27.08 per cent (around

4) said they didn't know how to do it.

In Class B, around 62.66% (an average of 13.16 students) already know how to classify the chemical elements using the Alternative Periodic Table. These students say that manipulating the cubes makes it easier to understand the division of chemical elements into groups and periods, but even so 3.17% said they didn't know how to do it, and 19.04% were in doubt when it came to solving the activities.

When the students had to indicate the group and period of the chemical elements based on the electronic distribution, it can be seen that the students who were able to solve the proposed questions were able to theoretically understand the organisation of the periodic table based on the studies of Mosley, 1914, who organised the chemical elements in ascending order of atomic number so that elements with similar physical and chemical properties were grouped together in the same column or group (RUSSEL, 1999).

Graphs 11 and 12 show the results obtained when the students identified the group and period of each chemical element from the electronic distribution.

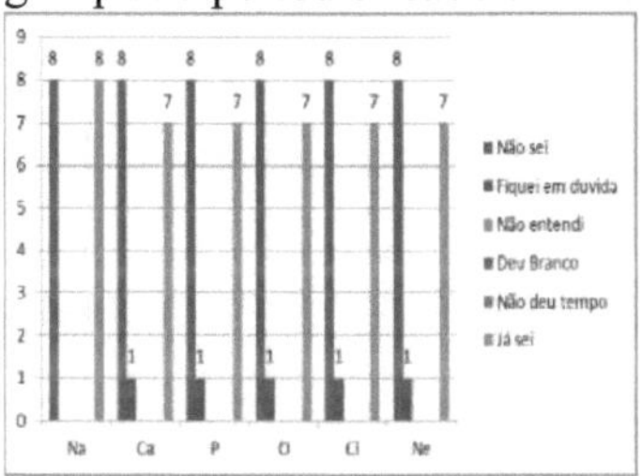

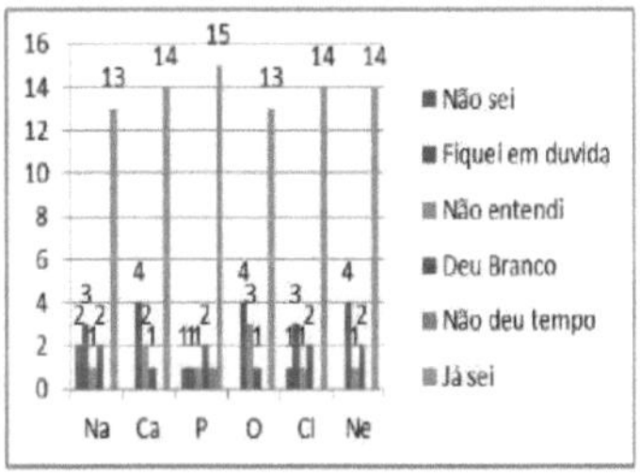

Graph 11: Classification of chemical elements into groups and periods based on their electronic distribution.

Graph 12: Classification of chemical elements into groups and periods based on their electronic distribution.

Source: Labour data, 2012.

In class A, 50% of the students were unable to solve the proposed exercises. This may be due to the fact that this special class has older students who migrated from the early primary school grades (5th, 6th, 7th) to the eighth grade with great difficulty in learning, as well as poor hygiene, low self esteem, diabetes, and students who were absent a lot due to illness. Many of these students have changed schools several times in the early grades and can't even read and write, have difficulties with interpretation, writing, Portuguese errors, maths difficulties in assimilating information, students with broken families, cases of aggression, relationship difficulties with classmates and teachers.

Approximately 44.37 per cent of the students said they understood the content well. The class has students who have many learning difficulties and, despite their limitations, they were interested in studying the periodic table, asked questions, dedicated themselves and didn't miss classes during the proposed activities.

Of the students who remained in doubt, around 5 per cent knew how to solve part of the exercise, [....] "they got lost in the middle of solving it", which could be overcome with a greater number of classes, after-school tutoring and the use of different teaching strategies.

Of Class B, 65.87% said they already knew how to solve the exercises, 15.07% were unsure and 3.17% couldn't do it. There were probably more correct answers because this class had manipulated the cubes representing the elements.

> From these exercises, students can verify the theory and justify their observations according to the periodic law, which states that the elements are listed sequentially in increasing order of atomic number, which is observed by the periodic repetition in their properties (RUSSEL,1994).

After analysing the electronic distribution for each element in the notebook, the students were able to compare their observations with the data contained below for each chemical element in the Alternative Table and also with the Traditional Table, confirming their results.

Graphs 13 and 14 show the results obtained in the exercises on formulating Lewis structures and differentiating between covalent and ionic chemical bonds.

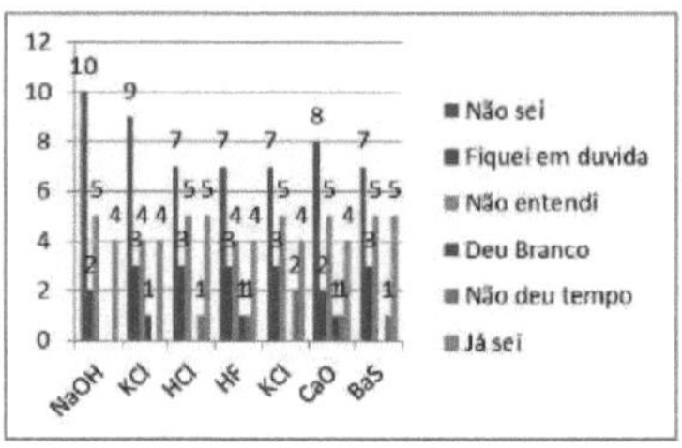

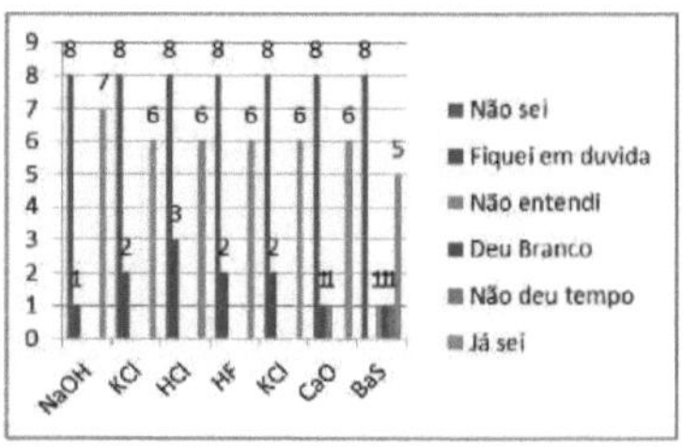

Graph13: Chemical Bonds: Ionic or Covalent. Class A.

Graph 14: Chemical Bonds: Ionic or Covalent. Class B.

Source: Labour data, 2012.

When these same students solved the exercise that asked them to represent Lewis chemical structures and classify chemical bonds as ionic or covalent, they realised that the degree of difficulty increased a lot, as most of the students replied that they didn't know or didn't understand, that they were in doubt, that they blanked and that they didn't have time.

In Class A, 49.10% said they didn't know how to do the exercises, 4.71% didn't understand, 16.96% were in doubt, and 27% already knew. In Class B, 38.09% didn't know how to do the exercises, 21.57% already knew how to do them and 8.16% were in doubt.

> The difficulty in proposing chemical structures and classifying them at any level of education is directly related to a lack of understanding of the organisation of the periodic table, electronic distribution and periodic properties. Although the chemical content is not intensively analysed in the eighth grade, fundamentals such as chemical bonds are covered through the study of the periodic table and the octet theory (CANTO, 2002).

Textbooks usually have an approach relating **chemical** bonds **to the octet rule, where the "tendency of atoms to lose or** gain electrons **to complete the octet" is the explanation for chemical stability** (an approach that can be given in the eighth grade), forgetting that it is the energy involved in the **situation that stabilises the compound and not the "will" of the atom to reach the** electronic configuration of the nearest noble gas. According to MORTIMER (1994), the **octet rule has its historical importance,** ***"after all, it's 45 years old"***, and is a useful procedure for predicting valence and compound formulae for elements in the periodic table, but it cannot be used as a dogma to explain bonds between atoms by accepting principles such as the energy variation involved in bond formation.

> MELO (2002) shows the importance of teaching chemistry through models and the unfavourable consequences for learning about chemical bonds when justified by the octet rule, such as: the macroscopic behaviour of ionic and molecular compounds is not **justified by the octet; it induces the student to think that the number of "bonds" between** the cation and the anion, for example, in the ionic bond, is limited to the number of electrons transferred; the sharing of a pair of electrons keeps the atoms together by reaching the octet, instead of sharing being seen as an energetically more favourable situation for the atoms than compared to the one where they are isolated; it makes it difficult to accept the reactivity of the noble gases; it does not justify the formation of compounds that have valence electrons from the 3rd energy level onwards; from the start of the bond, the electron pairs no longer belong to one nucleus or another, but to the molecule formed, so there is no point in using the term "dative **covalent bond**", for example.

The teacher's role is to make the necessary considerations at each level of Teaching the study of chemical bonds.

Figure 12 shows some substances that are incorrectly represented by the alkaloes from the Lewis structure using the alternative portable table.

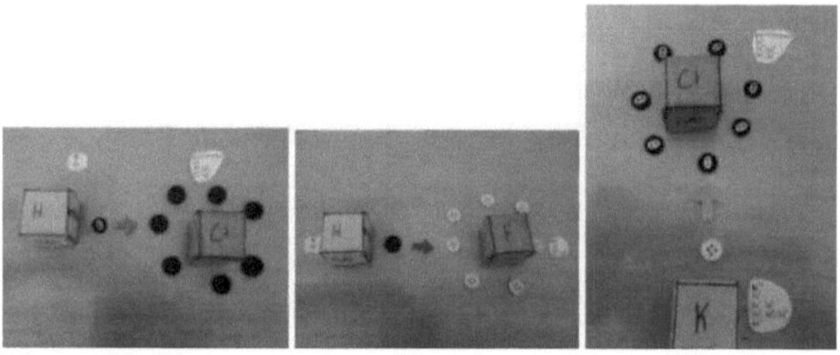

Figure 12: Representation of Lewis structures with Alternative Table

(HCl HF KCl)

Source: Labour data, 2012.

From the representations proposed for hydrochloric acid (HCl), hydrofluoric acid (HF) and potassium chloride (KCl), it can be seen that the students' greatest difficulty is in differentiating between the sharing and transfer of electrons in covalent and ionic bonds. This is probably because these students have not grasped the fundamental concepts (atomic properties such as atomic radius, ionisation energy and electroaffinity, for example) in the relationship and understanding of chemical bonds. In the representation of the structures of HF and HCl, they used arrows to indicate the bond, showing a transfer of electrons rather than sharing, since they are molecular substances formed from covalent bonds.

At the end of the activity, they were shown how to write molecular and ionic formulae, emphasising the differences. It was also pointed out that writing these formulas depends on certain rules and atomic properties and that these will be discussed in more detail in secondary school.

2.3.3 TEACHING CHEMISTRY: A PROPOSAL FOR THE STUDY OF PERIODIC TABLE IN SECONDARY SCHOOL

This work can also be aimed at high school chemistry students, provided it is suitably modified. For the chemistry teacher, it could be a support for an alternative development of their activities.

> According to Ausubel's theory (1890, p.10), which proposes that learning will only be meaningful for students if they are taught and encouraged to learn in this way, to the detriment of mechanical learning. In order to achieve better results, it is necessary for the

> concept map methodology and meaningful learning to occupy a place in schools from the first years of students' education, gradually being incorporated into everyday school life.
>
> Contextualising chemistry is not about promoting an artificial link between knowledge and the student's everyday life. It is not citing examples as an illustration **at the end of some content, but contextualising is proposing** "real problem **situations** and seeking the knowledge necessary to understand them and try to solve them." **(PCN+, p.93).**

In view of the above, it is necessary to practice more contextualised teaching, where the aim is to relate chemistry content to the daily lives of boys and girls, respecting the diversity of each one, with a view to training citizens and exercising their critical sense.

One of the aims of chemistry is for young people to recognise the value of science in the search for knowledge of objective reality and to insert it into everyday life. In order to achieve this goal, we try to work with contexts that are meaningful to the student and can lead them to learn, in an active process. It is believed that the student is involved not only intellectually, but also affectively. According to the new curriculum proposals (PCNs), this means educating for life.

Flowchart 1 shows which topics can be explored in secondary school using the alternative periodic table.

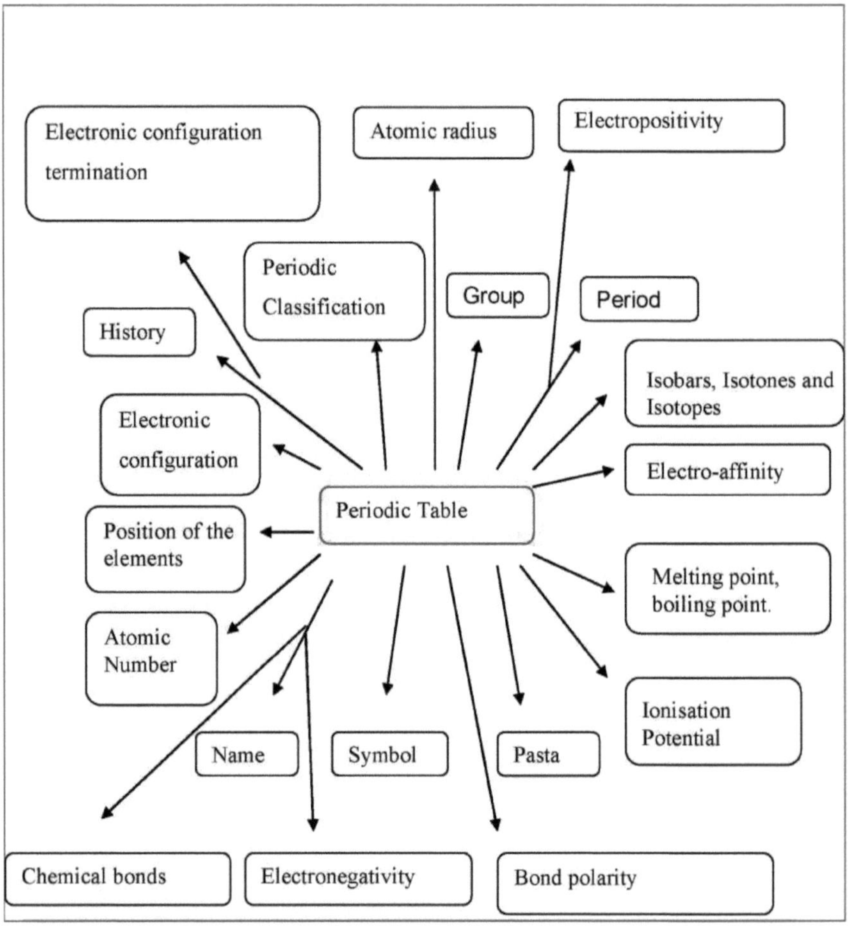

Flowchart 1: concept map for the study of the alternative table

Some aspects that can be explored using the Alternative Periodic Table:

C Knowledge of the atom and its fundamental particles (protons, neutrons and electrons).

S Symbology and Representation (chemical language).

F Formation and formulation of some everyday compounds.

L Reading and using the Periodic Table: periodic properties and chemical bonds.

CHAPTER 3

FINAL CONSIDERATIONS

Although the topic of paper recycling was not the main focus of the study, the selective collection of used paper and its recycling was significant for the development of the work, in the sense that it aroused the interest of the Flow Correction students, who have a lot of difficulty learning for various reasons, involving them more in the lessons. Recycling the paper and using it to produce an alternative table showed the students that it is possible to study science in a contextualised, creative way while taking care of the environment.

The results indicate that the use of alternative models for studying the periodic table is valid for examining the conceptual organisation that students have assimilated after an instructional period. The results are in line with Ausubel's Theory apud in Trindade, 2012, which proposes that learning will only be meaningful if students are predisposed to learning with understanding. In order to achieve better results, science/chemistry teachers need to use different teaching methodologies and strategies to stimulate interest and encourage meaningful learning. This type of initiative should take place at school from the first years of students' education, gradually being incorporated into everyday school life. Furthermore, it is important to emphasise that learning will only be meaningful if there is a conscious effort on the part of the learner. It can also be said that no matter how fertile the instruction, whether it takes in diverse contexts, seeks out new forms of assessment or even moves away from the traditional teaching model, without enthusiasm, motivation or intellectual curiosity on the part of the students, it is bound to produce timid results.

As the student is the mainstay or pillar of the entire teaching-learning process, their role is crucial to building meaningful and lasting learning (TRINDADE, 2012). It's worth pointing out that the diversified teaching contexts covered in this study, especially the activities in groups, were very well accepted by the students, which shows the learners' preference for situations that escape the monotonous routine of the blackboard and chalk, a reality in most public schools.

Thus, contextualised work and group activities that allow for manipulation, as was the case during the process of producing recyclable paper and the use of the alternative periodic table enables reflection, when combined appropriately, and can be an efficient resource for meaningful learning.

The use of the Alternative Periodic Table during science lessons has proved to be an effective strategic component in promoting student interest in the object of study, motivation, participation and appropriation of knowledge, stimulated by group work. However, these activities, in the search for knowledge, require the student to take active, enquiring, reflective, socialising and creative action, in total opposition to the submission, thoughtlessness and conditioning of the domineering pedagogy suggested by the traditional school, which is present in most schools and which should be encouraged during lessons by the teacher.

REFERENCES

ALMEIDA,S,C,E et al Centre for Exact and Natural Sciences/Department of Chemistry/ PROBEX - **VII Meeting on Teaching in Engineering Techniques instructional resources optimising the learning process,** 2013.

ALMEIDA, M.J.P.M. e RICON, A.E. **Divulgação científica e texto literário:** uma perspectiva cultural em aulas de Física. Caderno Catarinense de Ensino de Física, v. 10, p. 7-13, 1993

AUSUBEL, D. P. **Acquisition and retention of knowledge**: a cognitive perspective. Lisbon: Plátano, 2003. 219 p.

BAIRRO, junior, 2012 available at http://www.ijunior.te.pt/servlets/Bairro?P=Fazer&ID=76 accessed on 04/06/2012

BERBEL, N.N. **Problematisation and problem-based learning:** different terms or different paths? *Interface-Communicação, Saúde, Educação,* 1998-February; 2(2): 139-154.

BRAZIL. **National Education Guidelines and Bases Law,** Law No. 9.394, of 20 December 1996.

BRAZIL. Ministry of Health. **Secretariat for Labour Management and Education**. Department of Health Education Management. Project for the Professionalisation of Nursing Workers. *6 Pedagogical proposal:* the bases for action. Brasília: Ministry of Health, 2003e.

BRASILIA, 2011 **Pedagogical strategies for correcting age and grade distortion,** Government of the Federal District.

CAMPBELL, J. A. Atomic Size and the Periodic Table. **Journal of Chemical Education,** v. 66, n. 9, p. 739-740, dec. 1989

CARRASCOSA, J.; GIL-PÉREZ, D.; VILCHES, A. and VALDÉS, P. **Papel de la actividad experimental en la educación científica.** *Caderno Brasileiro de Ensino de Física*, v. 23, n. 2, p. 157-181, 2006.

CANTO, Eduardo Leite do. **Natural sciences:** learning from everyday life: grade 8· São Paulo: Moderna, 2002. 272 p. (Natural Sciences Collection: Learning from everyday life).

CACHAPUZ, A. F. C. (org.). **Perspectives on science teaching.** 1.ed. Porto: Centre for Studies in Science Education, 2000 (Teacher training - Science, 1).

Basic Education Curriculum Guidelines, 2008.

FAZENDA, I. C. A. **Integração e interdisciplinaridade no ensino brasileiro:** Efetividade ou ideologia. 5th ed. São Paulo: Loyola, 1979.

FAZENDA, Ivani C Arantes. **Integration and interdisciplinarity in Brazilian education:** effectiveness or ideology. 2. ed. São Paulo: Edições Loyola, 1992

FERNANDES, Maria Luiza Machado. **Teaching Chemistry and Everyday Life**. Curitiba: Ibpex, 2007.

FOUREZ, Gerard. **The Construction of Science**: Introduction to the Philosophy and

Ethics of Science. São Paulo: Editora da Universidade Estadual Paulista, 1995.

FREIRE, P. **Pedagogia da autonomia**. 29 ed. São Paulo: Paz e Terra, 2004

GIORDAN, M. **O papel da experimentação no ensino de ciências.** Química Nova na Escola, n. 10, p. 43-49, 1999.

GODOI F.D.A THIAGO, et al, **Periodic Table - A Super Trumpet for Primary and Secondary School Students**, 2009, Química Nova na Escola, Vol 32 N° 1, February 2010.

GONÇALVES,J.C. **Atomic Table:** a complete study of the periodic table. Curitiba: Ed Atômica, 2001, 1st ed.

GONÇALVES,J.C. **Tabela Periódica comentada,** Curitiba: Ed Atômica, 2003, 1st ed.

HERMES, R, A, Terezinha et al, p. 1, 2011)
Rescuing everyday chemistry in learning environments 31 EDEQ

LABURÚ, C.E. **Fundamentals for a captivating experiment.** Caderno Brasileiro de Ensino de Física, v. 23, n. 3, p. 382-404, 2006

LANAO,R,C.**Periodic table**: The periodic properties of the elements http://educacao.uol.com.br/disciplinas/Quimica/tabela-periodica-as-propriedades-periodicas-dos-elementos.htm , accessed on 02/01/2013

MAGALHÃES, Mariza. **Creative** Techniques **for Streamlining Chemistry Lessons.** Niterói: Muiraquitã, 2002

MELO, M.R. **Atomic structure and chemical bonds** - an approach for secondary education. Master's dissertation - State University of Campinas, São Paulo, 2002.

MENDES,P. **Brief history of the Periodic Table,** Published in the newspaper "Diário do Sul" on 11-12/3/2011

MIRANDA, KCL; BARROSO, M.G.T. **The contribution of Paulo Freire.** *Rev. Latino-Am Enfermagem,* 2004

MORTIMER, E. F.; MOL, G.; DUARTE, L.P. Octet rule and chemical bond theory in secondary education: dogma or science? **Química Nova**, v.17, n. 3. 1994.

Ministry of Education (MEC), Secretariat for Secondary and Technological Education (Semtec). **PCN + Ensino médio:** orientações educacionais complementares aos Parâmetros Curriculares Nacionais - Ciências da Natureza, Matemática e suas Tecnologias. Brasília: MEC/Semtec, 2002.

Ministry of Education (MEC), Secretariat for Secondary and Technological Education (Semtec). **National Curriculum Parameters for Secondary Education**. Brasília: MEC/Semtec, 1999.

NERICI, Imidio Giuseppe, **Metodologia do ensino**: uma introdução. 3rd ed. São Paulo: Atlas, 1989.

National Curriculum Parameters. 2. sciences. B823p Brazil. Secretariat of Fundamental Education. Available at: <http://portal.mec.gov.br/seb/arquivos/pdf/Ciências.pdf>. Accessed on: 03 May 2013.

PCNs - National Curriculum Parameters, BRAZIL 1998

RIBEIRO, F.C.P. **Teaching Chemistry**: alternative methodologies (SEDUC-PA)2010

RIOS, T. A. **Ethics and Interdisciplinarity**. In: FAZENDA, I. C. A. (Org.). A pesquisa em educação e as transformações do conhecimento. 2ª ed. Campinas: Papirus, 1997

RUSSELL, J. V. Using Games To Teach Chemistry. 2. CHeMoVEr Board Game.**Journal of Chemical Education**, v. 76, n. 4, p. 487, abr. 1999.

RUSSELL John B, 2nd edition volume 1, 1994, p. 298.

SANTANA, Olga. **Natural Sciences**. 8th year, 3rd edition, São Paulo, 2009 Editora Saraiva

SANTOS,C. MOREIRA,K. **Discovering Chemistry** 2011 Periodic Table available at http://cleytonlibelula.blogspot.com.br/2011/10/tabela-periodica.html, accessed on

13/02/2013

SILVA et.al (**Application of the didactic game Amarelinha da Química for students of the 1st year of high school, as a teaching-learning tool.**) 2012, available at (http://www.abq.org.br/simpequi/2012/trabalhos/162-12806.html) accessed on 29/05/2013 at 19:57

SILVA, H.C. and ALMEIDA, M.J.P.M. (Orgs.). **Languages, readings and the teaching of science**. Campinas: Mercado das Letras, 1998. p.121-130

SIMÕES, L. C. Interdisciplinarity at School. Available at: <http://www.peli.uem.br/artigoproducoes6.html>. Accessed on 07 March 2012. **CONTEXTUALISING CHEMISTRY TEACHING**: MOTIVATING SECONDARY SCHOOL STUDENTS

TASCA, Rodolfo Aureo.T181e **Structure of matter and periodic table in science teaching** for grade 8 - Alternative paths in chemistry teaching / Campinas, SP: [s.n], 2006.

TRINDADE, T. **Nanomaterials and the discovery of new worlds on the chemist's bench.** Chem. Nova vol.35 no.7 São Paulo 2012

MIX
Papier aus verantwortungsvollen Quellen
Paper from responsible sources
FSC® C105338

Printed by Books on Demand GmbH, Norderstedt / Germany